TELL ME HOW IT WORKS

HOW DOES ELECTRICITY WORK?

PHIL CORSO

PowerKiDS press
New York

Published in 2021 by The Rosen Publishing Group, Inc.
29 East 21st Street, New York, NY 10010

First Edition

Editor: Siyavush Saidian
Book Design: Reann Nye

Photo Credits: Cover zhangyang13576997233/Shutterstock.com; Series Art (gears) goodwin_x/Shutterstock.com; Series Art (newspaper) Here/Shutterstock.com; p. 5 Cavan Images/Cavan/Getty Images; p. 7 Hulton Archive/Getty Images; p. 8 Science & Society Picture Library/SSPL/Getty Images; p. 9 https://commons.wikimedia.org/wiki/File:Thomas_Edison2.jpg; p. 10 fokke baarssen/Shutterstock.com; p. 11 mycteria/Shutterstock.com; p. 13 SteveDF/E+/Getty Images; p. 15 Viktorus/Shutterstock.com; p. 16 Maskot/Getty Images; p. 17 Adnan NP/Shutterstock.com; p. 21 Hispanolistic/E+/Getty Images; p. 22 imacoconut/Shutterstock.com.

Cataloging-in-Publication Data

Names: Corso, Phil.
Title: How does electricity work? / Phil Corso.
Description: New York : PowerKids Press, 2021. | Series: Tell me how it works | Includes glossary and index.
Identifiers: ISBN 9781725318335 (pbk.) | ISBN 9781725318359 (library bound) | ISBN 9781725318342 (6pack)
Subjects: LCSH: Electricity–Juvenile literature.
Classification: LCC QC527.2 C66 2021 | DDC 537–dc23

Manufactured in the United States of America

CPSIA Compliance Information: Batch #CSPK20. For Further Information contact Rosen Publishing, New York, New York at 1-800-237-9932.

CONTENTS

POWER ON

You walk into a room and the first thing you do is flick the light switch. It might seem like an easy, simple thing. But what's going on behind the scenes? More than you might think!

Electricity is all about atoms, which are small **particles** that make up all matter. Inside those atoms are even smaller objects called **protons**, **electrons**, and **neutrons**. The way these particles act together is the secret to creating the electricity that powers our world.

The simple act of turning on the light sets off a chain of events that we call electricity.

WHO TURNED THE LIGHTS ON?

The year was 1752. A man named Benjamin Franklin looked outside, saw lightning, and went out to conduct an experiment.

He took a kite, got the string wet, put a metal key at the end, and then watched the kite float up into the storm above. Lightning struck and Franklin watched electricity travel down the string, giving him a shock.

This experiment was very dangerous, or unsafe—you should never try it—but it helped him learn about electricity.

TECH TALK

Many great minds after Franklin went on to help make electricity a more useful **utility**. Thomas Edison invented one of the first electric light bulbs in 1879.

This drawing shows Benjamin Franklin and his son William using a kite to experiment with electricity in 1752.

BRINGING IDEAS TO LIGHT

Thomas Edison invented the first electric light bulb that was cheap and easy to use. It passed an electrical current through metal **filament** wire inside the bulb, heating it up until it glowed. An electric current is the movement of electrons from a negatively charged area, or an area with more electrons, to a positively charged area, or an area with more protons. The filament was guarded from outside air by the glass around it, filled with a special gas that kept it working.

Thomas Edison's electric light bulb changed homes across the United States and the world.

FLIPPING THE SWITCH

Electricity is now everywhere. It flows through our homes. It powers the electric cars that get us where we need to go. Power plants use machines called generators to create, or generate, electricity. Power lines bring electricity to our homes.

TECH TALK

One way to make electricity is to burn natural resources, such as coal. San Francisco plans to generate, or create, half of its electricity through renewable resources—such as wind—in 2020.

Power lines are a common sight and are often overlooked, but they're responsible for providing communities with electricity.

The batteries in our cell phones store electric energy, or power, using chemicals, or matter that can be mixed with other matter to create changes. Batteries create their own electric currents using chemical energy. Over time, batteries must be recharged with electricity or recycled.

STEM ELECTRICITY

Electricity is a result of advancements in all the areas that make up STEM: science, technology, engineering, and math.

Voltage (a measurement of electrical energy), currents, resistance, power, and circuits are science concepts, or ideas. These things rely on the technology of lights, switches, motors, and more to create electricity. Making this technology requires engineering to design and test new advances. Math is used to figure out the voltage needed for electricity to flow correctly without being too strong or too weak.

TECH TALK

An electric circuit is a path for transmitting, or sending, an electric current. It has parts such as a battery that gives power to charged particles. Resistance turns electric power into heat due to an opposite current.

Circuit boards like this combine every STEM field to make sure electrical devices, including computers, work properly.

THE WORLD IS BUZZING

Electricity has changed the way we live for the better. Many of the improvements can be found in your house. In addition to lighting up your room and powering your TV, electricity can also give your heating and cooling systems power. A machine for heating called a furnace may run on electricity or fuel. Electricity usually powers air conditioners, which are machines that keep areas cool.

Electricity has also changed the way we eat. It has improved the way we cook by powering ovens and other kitchen appliances.

TECH TALK

Ovens can run on either gas or electricity. Gas ovens still use electricity to power features like a clock and an igniter, or the part that starts a fire.

Many of our homes use electricity from power plants that burn fossil fuels, or dead plants and animals. These plants release gases that trap the sun's heat, causing climate change, or long-term changes in Earth's average weather conditions.

MORE POWER!

Over the years, advancements in electricity have brought us into a new age of technology. The need for electricity has been growing with world populations. The energy **sector** has been quickly changing to conserve, or not waste, power whenever possible.

Solar panel technology is one of the most recent advancements in electricity. It allows homes to make their own energy and not rely on power plants that burn fossil fuels.

Energy-**efficient** light bulbs, for example, have become common in American households because they need less energy to work. Solar panels are also on the cutting edge of electricity, capturing energy from the sun to power homes, lighting, and even airplanes!

PLUGGED IN LIVING

Energy is made and used all over the world at different rates. China and the United States made up three-quarters of electricity production growth in 2018.

China led the world in production in 2018, with 7,092 TWh, or terawatt-hours, which is a measure of electrical energy. The United States had the number-two spot with 4,429 TWh.

China is first and the United States is second in consumption, mostly because these countries are large and use a lot of advanced technology.

TECH TALK

Most of the growth in global electricity production came from Asia in 2018. China accounted for most of that growth, coming in at nearly 60 percent of the global production growth.

GENERATING ELECTRICITY

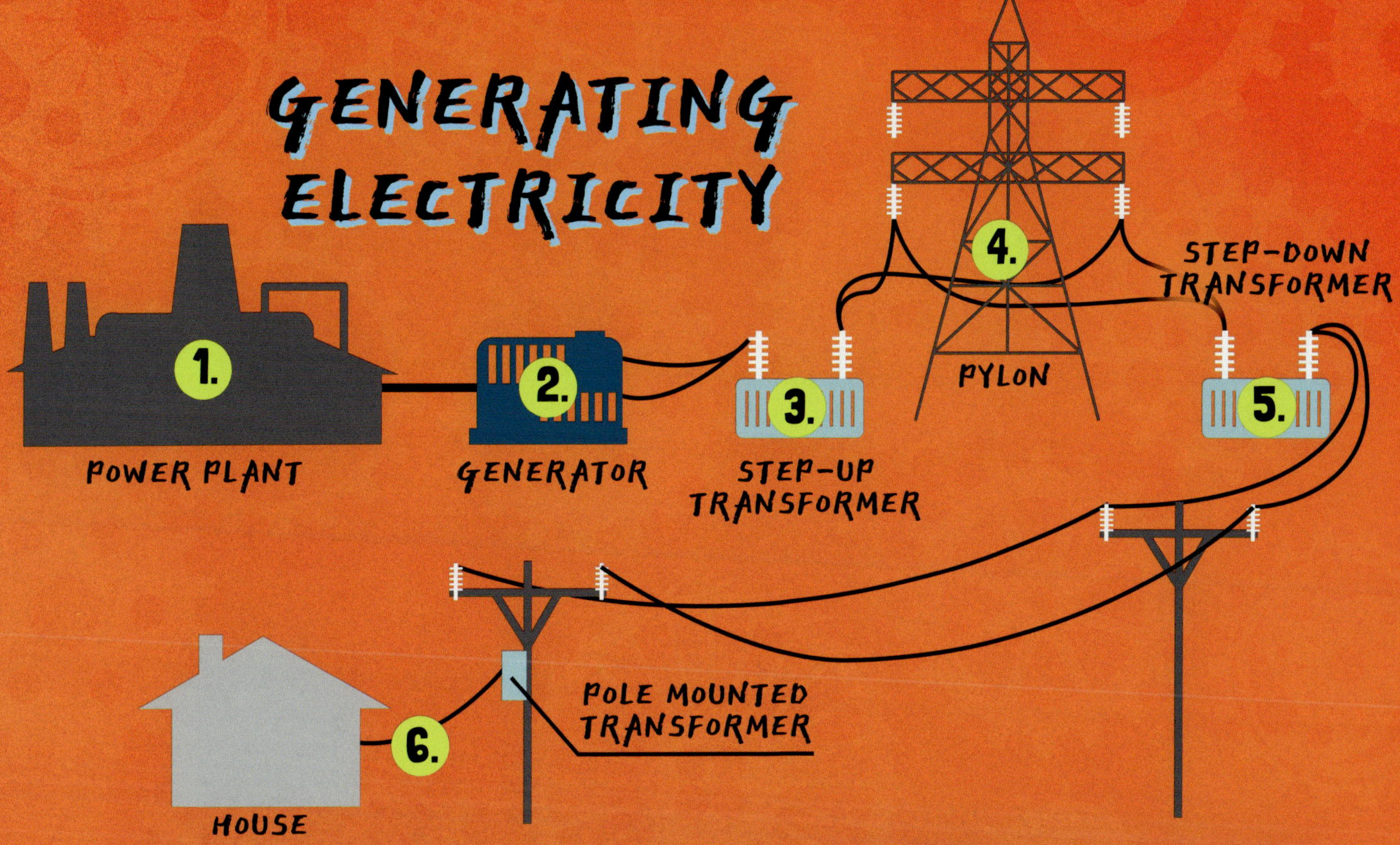

1. Fuel is burned to create steam that turns a **turbine**.
2. The turbine is connected to a machine called a generator that changes movement into electricity.
3. A step-up transformer boosts this energy to a high voltage because some of the energy can be lost as it travels through wires to homes.
4. Pylons are tall metal towers that take very high-voltage electricity in cables to where it needs to go.
5. A step-down transformer makes the energy a lower voltage that is safer for homes to use.
6. Electricity flows through overhead or underground cables into a home.

THEY'VE GOT THE POWER

The electricity business requires lots of hard work, and many different professions, or jobs, to keep the world lit up.

Electricians are in charge of electrical wiring and lighting systems. They're the ones to call when the lights go out, or when electrical wiring isn't working.

Power plant operators are also important. They keep up the equipment needed to generate electricity, including things like boilers, turbines, pumps, and fans. They work inside power plants to make sure power stays on.

TECH TALK

In the United States, there were more than 715,000 electricians in 2018. That same year, there were more than 50,000 working with power plants.

An electrician is one of many important jobs in the electricity business.

BRIGHT IDEAS

While electricity becomes more noticeable in modern society, so do improvements in creating and using it. The electricity of the coming years will be cleaner, require less energy, and be more **sustainable**.

Renewable energy is on the rise. In the United States, the ability to use wind to create electricity grew about 8 percent in 2018. Burning fossil fuels for electricity makes climate change and the deadly weather events it causes worse. Scientists say turning to renewable energy instead is our best shot at a brighter tomorrow!

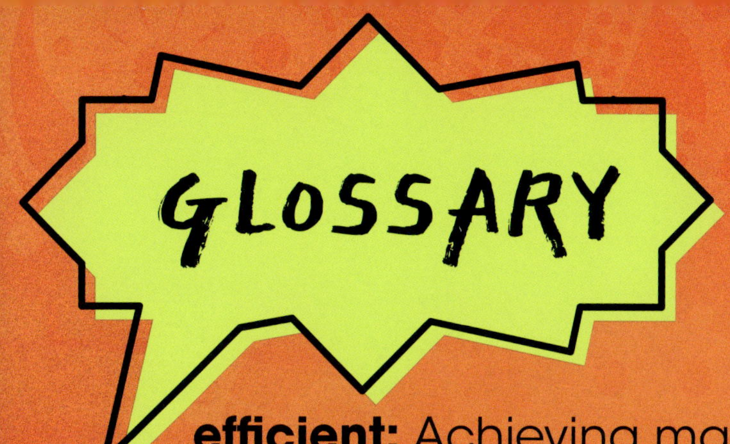

efficient: Achieving maximum productivity with minimum wasted effort.

electron: A stable subatomic particle with a charge of negative electricity.

filament: The thin wire in a light bulb.

neutron: A subatomic particle without an electric charge.

particle: A very small part of matter.

proton: A stable subatomic particle with a positive electric charge.

sector: An area or portion that is distinct from others.

sustainable: Able to last a long time.

turbine: An engine with blades that spin due to pressure from water, steam, or air.

utility: Something that is useful or beneficial.

Due to the changing nature of Internet links, PowerKids Press has developed an online list of websites related to the subject of this book. This site is updated regularly. Please use this link to access the list: www.powerkidslinks.com/tmhiw/electricity